United States Government Accountability Office

Report to Congressional Requesters

I0471927

January 2012

DEPARTMENT OF ENERGY

Additional Opportunities Exist to Streamline Support Functions at NNSA and Office of Science Sites

GAO

Accountability ★ Integrity ★ Reliability

GAO-12-255

GAO
Accountability * Integrity * Reliability

Highlights

Highlights of GAO-12-255, a report to congressional requesters

DEPARTMENT OF ENERGY

Additional Opportunities Exist to Streamline Support Functions at NNSA and Office of Science Sites

Why GAO Did This Study

The Department of Energy (DOE) spends 90 percent of its annual budget—which totaled $27 billion in fiscal year 2011—on the contractors that carry out its diverse missions and manage its sites. These management and operating contractors also provide sites' support functions, such as procuring goods, managing human resources, and maintaining facilities. With a unique contractor at each site, support functions have traditionally been managed in a decentralized, or fragmented, manner. In light of today's pressures to trim budgets and find efficiencies, GAO was asked to review support functions at the 17 National Nuclear Security Administration (NNSA) and Office of Science sites and determine (1) the costs of providing support functions for fiscal years 2007 through 2011; (2) efforts undertaken during that period to streamline sites' support functions, as well as additional opportunities and challenges, if any; and (3) the extent to which cost savings from streamlining efforts can be quantified. GAO reviewed data and documents and spoke with DOE, NNSA, and Science officials and with contractors at eight sites—the four largest by budget from NNSA and Science.

What GAO Recommends

GAO recommends that DOE (1) fully implement a quality control system for cost data on sites' support functions, (2) ensure that all appropriate streamlining steps are being taken at the 17 sites and that challenges are addressed, and (3) clarify guidance on estimating cost savings from streamlining efforts. DOE agreed with the recommendations.

View GAO-12-255. For more information, contact Gene Aloise at (202) 512-3841 or aloisee@gao.gov.

What GAO Found

Support function costs at NNSA and Science sites for fiscal years 2007 through 2011 are not fully known because DOE changed its data collection approach beginning in 2010 to improve its data and, as a result, does not have complete and comparable cost data for all years. In fiscal years 2007 through 2009, total support costs for NNSA and Science sites grew from $5 billion to about $5.5 billion (nominal dollars). Costs for fiscal year 2010 are unknown because DOE was pilot-testing its new reporting system and only collected data from some sites. For fiscal year 2011, the data are more complete, but changes to DOE's definitions for support functions make it difficult to compare costs across all years. DOE has taken some steps to ensure the quality of the data in its new system and plans to fully implement a quality control process, such as peer reviews, to ensure data can be compared across sites, but has not yet done so.

DOE and contractors have undertaken various efforts since 2007 to streamline and improve the efficiency of sites' support functions. Streamlining efforts reported by officials from DOE and the eight NNSA and Science sites GAO reviewed focused mainly on procurement; human resources, including employee benefits; and facilities and infrastructure. Some efforts were part of larger initiatives involving multiple sites, while others were initiated at the site level. To streamline procurement and leverage the buying power of multiple sites, for example, NNSA began operating a central Supply Chain Management Center to negotiate with vendors for lower prices on goods and services, such as laboratory supplies and equipment. To streamline human resources, contractor officials from the eight NNSA and Science sites reported automating various processes, such as for hiring and training employees. Furthermore, DOE and contractors identified opportunities to expand these efforts and undertake new ones but also identified challenges to further streamlining. In August 2010, for example, the Deputy Secretary of Energy cited further opportunities to leverage DOE and sites' buying power through a more centralized, and less fragmented, approach. Similarly, NNSA is considering centralizing certain human resource tasks at its sites, currently provided by individual contractors. DOE and contractor officials, however, said that centralizing functions can be challenging.

DOE and its contractors have estimated savings for some streamlining efforts, particularly in procurement, but it is difficult to compare or quantify total savings across sites because DOE's guidance for estimating savings is unclear and the methods used to estimate savings vary. For example, one laboratory estimated a $9 million savings from a software purchase in 2010 using its preferred estimation method. By other methods used elsewhere in DOE, however, the site estimated that its savings could have been as high as $35 million. DOE recently issued guidance on acceptable methods for estimating procurement cost savings, but the guidance is unclear and could lead to widely varying savings estimates. The guidance identifies some estimation methods that sites can use—such as comparing the price paid for goods or services with a previous price—but does not specify which methods are preferred when multiple options are available. Furthermore, the guidance allows sites to use any other methods approved by DOE officials. For support functions other than procurement, sites also have flexibility in cost savings estimation methods, potentially leading to widely varying estimates for similar efforts to streamline these functions.

_____ United States Government Accountability Office

Contents

United States Government Accountability Office
Washington, DC 20548

January 31, 2012

The Honorable Fred Upton
Chairman
The Honorable Henry A. Waxman
Ranking Member
Committee on Energy and Commerce
House of Representatives

The Honorable Cliff Stearns
Chairman
The Honorable Diana DeGette
Ranking Member
Subcommittee on Oversight and Investigations
Committee on Energy and Commerce
House of Representatives

The Honorable John D. Dingell
House of Representatives

The Department of Energy (DOE) spends a major portion of its annual budget—which totaled $27 billion in fiscal year 2011—to carry out groundbreaking scientific research and technology development to increase knowledge about fundamental physics, provide efficient and secure energy, and ensure the safety and reliability of the nation's nuclear weapons stockpile. This research and development work is performed by contractors—corporations, universities, and others—that manage and operate the 7 national laboratories and nuclear production and testing sites overseen by the National Nuclear Security Administration (NNSA),[1] a separately organized agency within DOE, and the 10 national laboratories overseen by DOE's Office of Science. With DOE oversight, these management and operating (M&O) contractors also organize and carry out the support functions at these 17 sites, such as procuring needed goods and services; recruiting and hiring workers; managing health and retirement benefits; maintaining facilities and infrastructure; and providing day-to-day accounting, information technology, security,

---

[1]Congress created NNSA as a semiautonomous agency within the Department of Energy in 1999 (Title 32 of the National Defense Authorization Act for Fiscal Year 2000, Pub. L. No. 106-65, § 3201 et seq.).

and other support functions.[2] Because each NNSA or Science site has historically had its own unique M&O contractor—as part of DOE's long-standing model for conducting research and nuclear production at multiple locations—the sites have also differed in how support functions are organized and carried out.[3]

DOE reimburses its M&O contractors for costs incurred in carrying out the department's missions and providing sites' support functions. These include costs that can be directly identified with a specific DOE program (direct costs) and costs that support multiple programs (indirect costs). Federal Cost Accounting Standards and federal regulations allow DOE's M&O contractors flexibility in how they classify incurred costs as direct or indirect.[4] Because sites classify these costs differently, in the mid-1990s, DOE's Chief Financial Officer began requiring M&O contractors at the sites to report on 22 categories of these costs—known as functional support costs—to provide more comparable data on the costs of sites' support functions.

We have previously reported on DOE's support costs and related issues. In September 2005, we reported that definitions for some of the functional support costs were unclear and that M&O contractors' reporting of these

---

[2]DOE and its M&O contractors' relationships are defined in federal and DOE acquisition regulations and in DOE's M&O contracts. M&O contracts are agreements under which the government contracts for the operation, maintenance, or support, on its behalf, of a government owned or controlled research, development, special production, or testing establishment wholly or principally devoted to one or more major programs of the contracting federal agency. Federal Acquisition Regulation, 48 C.F.R. § 17.601.

[3]In a July 2011 draft solicitation to industry, DOE and NNSA proposed altering this long-standing approach by having a single contractor manage and operate two of NNSA's nuclear production sites that have historically had their own M&O contractors. DOE and NNSA estimated that the new approach would save around $895 million (nominal dollars), largely through efficiency gains and other improvements to the sites' business systems and support functions. In September 2011, however, we reported that the anticipated cost savings were uncertain. See GAO, *Modernizing the Nuclear Security Enterprise: The National Nuclear Security Administration's Proposed Acquisition Strategy Needs Further Clarification and Assessment*, GAO-11-848 (Washington, D.C.: Sept. 20, 2011).

[4]Cost Accounting Standards are promulgated under chapter 99 of Title 48, U.S. Code of Federal Regulations, by the U.S. Cost Accounting Standards Board—a statutorily established board (41 U.S.C. § 1501) within the Office of Management and Budget's Office of Federal Procurement Policy. The standards are mandatory for use by all executive agencies and federal contractors in estimating, accumulating, and reporting costs.

costs was inconsistent. We recommended that DOE take further actions to improve the comparability of its data by clarifying its definitions for its support costs.[5] In 2010, DOE replaced its functional support costs with a new system called Institutional Cost Reporting, which was pilot-tested that year and fully implemented in 2011. Also in the 2005 report, and an April 2004 report,[6] we recommended that DOE take actions to manage the long-term cost growth in certain support functions, such as facility maintenance, as well as pension or other costs at sites. Since that time, DOE has taken actions to control these costs, but some of them have continued to grow. For example, in April 2011 we reported that DOE reimbursed M&O contractors departmentwide for $750 million in pension costs in fiscal year 2009—more than double the amount reimbursed in fiscal year 2008—following financial market declines.[7]

Against the backdrop of growing federal deficits and uncertainty over future federal budgets, DOE and its M&O contractors at NNSA and Science sites have been evaluating areas that could be streamlined or provide cost savings. In this context you asked us to examine support functions at NNSA and Science sites. Our objectives for this report were to examine (1) the costs of providing support functions at NNSA and Science sites for fiscal years 2007 through 2011; (2) efforts undertaken during that period to streamline sites' support functions and additional streamlining opportunities and implementation challenges, if any; and (3) the extent to which cost savings from streamlining efforts can be quantified.

To address the first objective, we analyzed DOE's data on support function costs at the 17 NNSA and Science sites for fiscal years 2007 through 2011 and interviewed DOE officials who oversee these cost data for DOE's Office of the Chief Financial Officer. We took steps to assess the reliability of the cost data, including interviewing representatives from

---

[5]GAO, *Department of Energy: Additional Opportunities Exist for Reducing Laboratory Contractors' Support Costs*, GAO-05-897 (Washington, D.C.: Sept. 9, 2005).

[6]GAO, *Department of Energy: Certain Postretirement Benefits for Contractor Employees Are Unfunded and Program Oversight Could Be Improved*, GAO-04-539 (Washington, D.C.: Apr. 15, 2004).

[7]See GAO, *Department of Energy: Progress Made Overseeing the Costs of Contractor Postretirement Benefits, but Additional Actions Could Help Address Challenges*, GAO-11-378 (Washington, D.C.: Apr. 29, 2011).

GAO-12-255 Support Functions at DOE Sites

the chief financial officers' organizations in DOE, NNSA, Science, and M&O contractors. We noted the limitations of these data in our report but found the data sufficiently reliable for our purposes. We reviewed federal Cost Accounting Standards and federal and DOE acquisition regulations for requirements on reporting support function costs. We spoke with M&O contractor officials responsible for financial management at a nonprobability sample of 8 sites—the 4 largest (by budget) NNSA sites and 4 largest Science sites—and discussed trends in the sites' support function costs since 2007.[8] Because a nonprobability sample is not generalizable, what we found at these 8 sites cannot be projected to all 17 sites; however, the sites provide examples of issues related to management of support functions.[9] We visited 3 of these sites—Los Alamos, Sandia, and Pacific Northwest national laboratories—and contacted the others by phone.[10] To address the second and third objectives, we reviewed DOE's policies on procurement, human resources, facility maintenance, and other support functions. We also spoke with headquarters and field-based officials who oversee support functions for DOE, NNSA, and Science about DOE's policies and efforts to oversee M&O contractors' performance in carrying out and streamlining support functions at sites, as well as additional streamlining opportunities and challenges. We also reviewed studies, cost reports, strategic plans, and other documentation on recent or proposed efforts to streamline NNSA and Science sites' support functions. We spoke with M&O contractor officials who plan and oversee the 8 sites' support functions and discussed their sites' streamlining efforts since 2007, as well as any cost savings from those efforts. We reviewed documentation on sites' recent and planned streamlining efforts and associated cost savings. We also discussed additional streamlining opportunities and any

---

[8]The four largest NNSA sites—Los Alamos, Lawrence Livermore, and Sandia national laboratories and the Y-12 National Security Complex—accounted for about 79 percent of the budget for NNSA's sites in fiscal year 2010. The four largest Science sites—Oak Ridge, Brookhaven, Pacific Northwest, and Lawrence Berkeley national laboratories—accounted for about 66 percent of the budget for that office's sites in that year.

[9]Furthermore, some of these issues may be relevant for sites overseen by other DOE organizations, such as the Office of Environmental Management. Our scope, however, did not include DOE organizations other than NNSA and Science.

[10]We visited both NNSA and Science sites, as well as larger and smaller sites among the 8 in our nonprobability sample. Furthermore, Pacific Northwest National Laboratory was involved in developing DOE's new Institutional Cost Reporting system and the associated pilot test.

related challenges. Furthermore, we collected, through document requests and interviews with DOE and contractor officials at sites in our sample, information on how cost savings were estimated. The amount and level of detail of this information varied greatly across streamlining efforts. Because it was not the purpose of this report to assess the anticipated or actual success of efficiency efforts and because the amount and quality of data on how estimated and actual savings were determined varied so much across efforts, we did not attempt to independently verify the reliability of these data or estimates. As a result, data on reported estimated or actual cost savings and efficiencies are of undetermined reliability.

We conducted this performance audit from December 2010 through January 2012 in accordance with generally accepted government auditing standards. Those standards require that we plan and perform the audit to obtain sufficient, appropriate evidence to provide a reasonable basis for our findings and conclusions based on our audit objectives. We believe that the evidence obtained provides a reasonable basis for our findings and conclusions based on our audit objectives.

Background

DOE is responsible for a diverse set of missions, including nuclear security, energy research, and environmental cleanup. These missions are managed by various organizations within DOE and largely carried out by M&O contractors at DOE sites. NNSA and Science are among the largest (by budget) of these DOE organizations, overseeing important missions at 17 sites.[11] Specifically:

- With a $10.5 billion budget in fiscal year 2011—nearly 40 percent of DOE's total budget—NNSA is responsible for providing the United

---

[11]In addition, some of the work for NNSA and Science missions is conducted by M&O contractors at sites overseen by other DOE organizations. For example, NNSA funds nuclear production and reprocessing work conducted by the M&O contractor at DOE's Savannah River Site in South Carolina, a site that is primarily overseen by DOE's Office of Environmental Management. Furthermore, some of the work carried out at NNSA and Science sites is funded by non-DOE entities, including other federal agencies or private firms. This work, known as "work for others," can comprise very little of the work at NNSA and Science sites, or in the case of Sandia National Laboratories in New Mexico and California, over 40 percent of a site's research budget in some years. In addition, NNSA and Science fund work at non-DOE organizations, such as universities.

States with safe, secure, and reliable nuclear weapons in the absence of underground nuclear testing and maintaining core competencies in nuclear weapons science, technology, and engineering. NNSA's 7 sites, including 3 national laboratories and 4 nuclear and production and testing sites support these activities (see fig. 1).

- With a $4.9 billion budget in fiscal year 2011—18 percent of DOE's total budget—Science has been the nation's single largest funding source for basic research in the physical sciences, supporting research in energy sciences, advanced scientific computing, and other fields. Science funds research at its 10 national laboratories, which also house cutting-edge scientific facilities and equipment, ranging from high-performance computers to ultrabright X-ray sources for investigating fundamental properties of materials. These resources are often made available, on a temporary basis, to members of the broader scientific community outside of DOE for their own research, sometimes in collaboration with laboratory staff.

Figure 1: NNSA's Laboratories and Nuclear Production and Testing Sites and Science's Laboratories

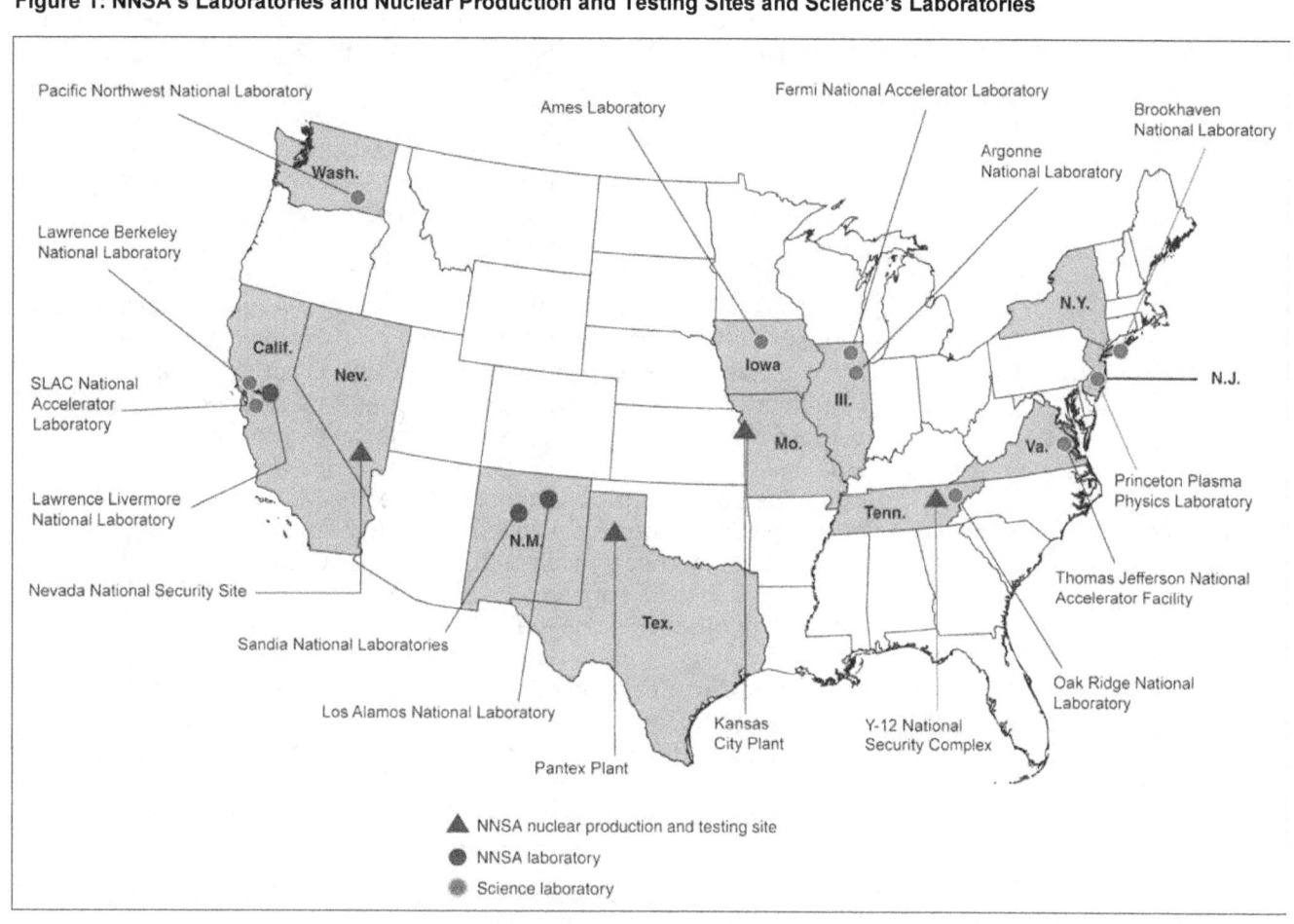

Sources: DOE; Map Resources (map).

Note: Sandia National Laboratories has locations in New Mexico and California. Only its primary location in New Mexico is depicted. Both locations are considered to be part of a single site among the 7 NNSA laboratory and nuclear production and testing sites.

Under DOE's long-standing model of having unique M&O contractors at each site, management of its sites has historically been decentralized and, thus, fragmented. Since the Manhattan Project produced the first atomic bomb during World War II, DOE and its predecessor agencies have depended on the expertise of private firms, universities, and others to carry out research and development work and efficiently operate the facilities necessary for the nation's nuclear defense.[12] DOE's relationship with these entities has been formalized over the years through its M&O contracts—contracts of a special type that give DOE's contractors unique responsibility to carry out major portions of DOE's missions and apply their scientific, technical, and management expertise. Currently, DOE spends 90 percent of its annual budget on M&O contracts, making it the largest non-Department of Defense contracting agency in the government. The contractors at DOE's 17 NNSA and Science sites have operated under DOE's direction and oversight but largely independently of one another.[13] Furthermore, M&Os are set up as separate entities with their own missions, parent organizations, and organizational structures. For example, the M&O contractor at Science's Oak Ridge National Laboratory is a private, not-for-profit company, established for the sole purpose of managing and operating that laboratory for DOE. Formed in 2000 as a limited liability partnership between the University of Tennessee and Battelle Memorial Institute, the M&O contractor organization is uniquely set up to provide that laboratory's scientific research and necessary support functions. M&O contractors typically allocate the costs of support functions across the site and charge research or other divisions that benefit from those functions.

Requirements for providing support functions at sites are set out in federal and DOE acquisition regulations, DOE policies, and M&O contracts. In particular, M&O contracts define performance requirements for the support functions provided at each site. These range from requirements that apply to contractors at all 17 sites—such as a

---

[12]The Manhattan Project, under the U.S. Army Corps of Engineers, maintained control over American atomic weapons research and production until the formation of the United States Atomic Energy Commission in 1946. The commission became the Energy Research and Development Administration in 1974, which in turn became DOE in 1977.

[13]M&O contractors in NNSA and Science interact regularly to share best practices or coalesce around topics of mutual interest. Examples include the National Laboratory Director's Council—in which representatives from NNSA, Science, and other DOE laboratories meet regularly to coordinate around issues and concerns of broad interest to laboratory managers—and DOE's newly formed Contractor Integrated Supply Chain Management Council, in which DOE and contractor officials will discuss issues related to procurement. In addition, sites' mission work is sometimes interrelated. For example, programs to refurbish nuclear weapons typically rely on NNSA's laboratories to design the refurbished weapon, which NNSA's production and testing sites will manufacture. Similarly, experts from various NNSA and Science laboratories have coordinated some of their sites' research in areas such as materials science and high-performance computing.

requirement that M&O contractors' systems of accounts comply with generally accepted accounting principles—to site-specific requirements, such as those pertaining to security at sites that house special nuclear material. Some M&O contracts also include provisions encouraging contractors to reduce sites' support function costs. For example, M&O contracts at some sites allow the contractors there to redirect—to mission work or contractor-directed research—dollars saved through reductions in support function costs. Also, contracts may include annual performance goals for cost savings that are tied to the contractors' annual performance fees, which are monies paid to them based on their annual performance.

Various headquarters and field-based organizations within DOE, NNSA, and Science develop policies or oversee M&O contractors' performance in providing support functions at sites. DOE, NNSA, and Science chief financial officers, for example, oversee requirements for contractor reporting on sites' support function costs. Other offices oversee particular support functions, such as procurement, facilities and infrastructure, and human resources. DOE's Office of Procurement and Assistance Management and NNSA's Office of Acquisition Management, for example, establish policies and oversee procurement activities at NNSA and Science sites. In addition, NNSA and Science site offices, colocated with the 17 sites, conduct day-to-day oversight of the M&O contractors and evaluate the contractors' performance in carrying out the sites' missions and providing support functions.

Sites' Support Costs since 2007 Are Not Fully Known

Support Cost Data for 2007 through 2011 Are Not Complete or Comparable because DOE Changed Its Data Collection Approach

The costs of support functions at NNSA and Science sites for fiscal years 2007 through 2011 are not fully known because DOE changed its data collection approach beginning in 2010 to provide improved data and, as a result, does not have complete and comparable year-over-year cost data for all 5 years. For fiscal years 2007 through 2009, total support costs for 16 of the 17 NNSA and Science sites grew from $5 billion to about $5.5 billion (nominal dollars) and generally accounted for about 40 percent of the sites' annual budgets for those years, according to DOE's data.[14] The proportion of total costs that support costs represented, however, differed between NNSA and Science sites in those years. For the NNSA sites, DOE's data show that support costs made up from 43 to 45 percent of sites' annual budgets for fiscal years 2007 through 2009. For the Science sites, the data show that support costs made up from 32 to 33 percent of the sites' annual budgets during those years. Differences in missions may account in part for NNSA's somewhat higher support costs. In particular, NNSA sites are more likely than Science sites to house nuclear material and classified information, which can result in higher security, training, or other costs. We did not, however, attempt to analyze costs for specific support functions. According to a DOE report on its sites' support function costs and previous GAO work,[15] the data are appropriate for understanding sites' support function costs in aggregate but not for comparing costs of sites' individual support functions.

In fiscal year 2010, DOE changed its data collection approach to improve its ability to oversee sites' support function costs, according to DOE officials, and, as a result, it does not have complete and comparable data for all fiscal years from 2007 through 2011.[16] DOE does not have complete data for 2010 because DOE and some of its contractors were pilot-testing the new system, and only 11 of the 17 NNSA and Science sites provided support cost data for that year. The data for 2010 are also incomplete because—unlike previous years' data—the sites only reported on the indirect costs of their support functions, leaving out direct costs, which could account for potentially large portions of their support costs.

---

[14]Over the same period, the sites' total annual support function costs increased from about $5.0 billion to about $5.3 billion in constant 2007 dollars. DOE's data did not include information for one Science site, the Thomas Jefferson National Accelerator Laboratory. Excluding that site, however, would likely have had little impact on the total support costs for fiscal years 2007 through 2009 because the site's budget ranged from about $94 million to about $130 million and represented only about 1 percent of the total budget in those years for all 17 sites.

[15]GAO-11-848.

[16]According to DOE's Office of the Chief Financial Officer, additional insight into sites' support function costs was needed in order to improve comparability with DOE's Standard Accounting and Reporting System, which came online after DOE initially started collecting support cost data. Also, the support cost data under the old reporting system did not differentiate between direct and indirect support costs. Furthermore, costs were reported annually under the old system, but are being reported quarterly under the new reporting system.

For example, officials from Los Alamos National Laboratory in New Mexico told us that because just 15 percent of that site's $145 million in annual security costs are indirect, reporting only the indirect costs would exclude $125 million in direct costs, or more than 85 percent, of the site's total security costs.

The data were more complete for fiscal year 2011—DOE's first year for implementing the new system after the pilot test; however, the 2010 and 2011 data for individual support functions are not comparable with older data because the definitions for some support functions changed from the old system to the new one. The new system retained many of the support function categories of the old system, but it changed the definitions for some categories and added new ones, such as cyber security, technology transfer, and internal audit. For example, the new system retained the safeguards and security category of the old system; however, it changed the definition for that category to exclude cyber security costs, which were made into a new cyber security category. Furthermore, DOE's definition for the new cyber security category identified specific activities, such as purchasing and maintaining security software, that are not described under the old system. DOE also added a technology transfer category to the new reporting system, which includes costs from various categories under the old system. Because the definitions changed in these and other instances, costs that were in one category under the old system were defined differently under the new system. As a result, costs of individual support functions from the two systems cannot be readily compared.

DOE Has Not Yet Taken All Planned Steps to Ensure Data Are Complete and Comparable

DOE has taken some steps to ensure the completeness of support cost data under its new data collection approach, but officials and documents described the need for additional steps to ensure that support cost data collected under the new approach are comparable across DOE sites. DOE has taken steps to ensure the completeness of data under the new approach, which requires M&O contractors to verify they are capturing all relevant costs by comparing their support costs with cost information from DOE's central accounting system.[17] DOE's Office of the Chief Financial Officer, however, plans to take additional steps, such as peer reviews, to

---

[17]Support function costs are compared with data from DOE's Standard Accounting and Reporting System. Specifically, sites check that their combined direct and indirect costs are equal to the total costs in that system.

ensure that the new data are comparable. Under the previous approach for collecting support function costs, DOE used peer reviews to help ensure that sites identified the relevant costs and assigned them to the correct categories. Specifically, the accuracy of sites' reporting—including whether the relevant costs were being included in the correct categories—was reviewed every 3 years by representatives from other M&O contractors. The reviews found instances of sites classifying costs incorrectly, which were reported to DOE and corrected by the contractor. In fiscal year 2011 guidance for the new system from the Office of the Chief Financial Officer, DOE stated that it planned to implement peer reviews by the end of fiscal year 2011. In September 2011, officials responsible for DOE's support cost data told us that the peer reviews had not been implemented because of other priorities, but more recently, the department said it plans to implement peer reviews in fiscal year 2012. Until these reviews have been implemented, it is difficult to know if the data collected under the new system are reliable or useful for comparing sites' support costs.

Complete and comparable data on support costs at NNSA and Science sites are not available for all years since 2007, but M&O contractor officials at the 8 sites we reviewed discussed trends in support function costs at each individual site. Because each site is different and M&O contractors have discretion in how they classify costs, contractors' own data systems may capture support costs differently than DOE's reporting systems discussed earlier. Contractor officials responsible for financial management at most of the 8 sites told us that support function and other costs at their sites, including pensions, had increased overall for fiscal years 2007 through 2011. These overall costs increased during that period, but some types of support costs generally decreased or remained stable relative to the total costs of managing and operating the sites, according to these officials. The costs that decreased generally included human resources administration, financial management systems, and other support costs that some officials said are easier to control. Contractor officials at the 8 sites said they used reductions in some support costs to help offset increases in support costs that can be more difficult to control, such as pensions and facility costs. For example, on the basis of information provided by Sandia National Laboratories, the site expects to contribute nearly $2 billion to its pension plan over the next 10 years, $400 million of which officials expect to fund through reductions in that site's support function costs. M&O contractor officials responsible for financial management at Brookhaven National Laboratory told us they plan to leverage the funds resulting from decreases in some support function costs in order to help fund ongoing efforts to upgrade the site's aging facilities and infrastructure, which had grown more costly to maintain. Specifically, rather than lowering the rates charged for the support functions, the M&O officials opted to continue charging the same rates and then use the additional funds to help pay for the upgrades.

DOE and Site Contractors Have Been Streamlining Support Functions and Identifying Additional Opportunities, as well as Challenges

Streamlining since 2007 Has Mainly Focused on Procurement, Human Resources, and Facilities and Infrastructure

DOE and M&O contractors have undertaken various efforts since 2007 to improve the efficiency of, or streamline, sites' support functions. Streamlining efforts that officials reported were focused mainly on three broad areas: (1) procurement; (2) human resources, including employee benefits; and (3) facilities and infrastructure.[18] Some streamlining efforts were part of larger initiatives involving multiple sites from among the 17 NNSA and Science sites, while others were initiated at the site level.

To streamline procurement, DOE and M&O contractors at NNSA and Science sites said that they have undertaken various efforts to obtain better pricing on goods and services and make their procurement processes more efficient. Some of the efforts were aimed at reducing fragmentation among sites by using a more centralized approach. For example:

- To better leverage its 7 sites' purchasing power, in 2007 NNSA began operating its central Supply Chain Management Center. According to NNSA, this center applies "strategic sourcing" techniques, coordinated

---

[18]The discussion below provides examples of streamlining efforts but is not meant to be comprehensive.

through a central organization, that private companies and, more recently, government agencies have used to improve efficiency and effectiveness of procurement.[19] The center aggregates and analyzes NNSA sites' procurement spending data to identify opportunities to coordinate sites' purchases and negotiate better prices for goods and services. For example, center officials conducted one analysis that revealed that the 7 sites were purchasing most of their laboratory supplies and equipment from the same set of 38 vendors through individual contracts negotiated by each site. The center was able to negotiate a single contract for all the sites, resulting in an estimated savings of $22 million over the contract's 3-year term, according to the M&O contractor official who manages the center. The official said the center has negotiated agreements for other goods and services, such as electrical supplies and equipment. Furthermore, NNSA's center supplements its own strategic sourcing agreements with ones negotiated by DOE's Integrated Contractor Purchasing Team for goods such as office supplies and personal computers. These agreements are similar to agreements negotiated by NNSA's center but, unlike the NNSA agreements, are available to M&O contractors across DOE. In addition, NNSA's central Supply Chain Management Center provides automated tools, including electronic catalogs, to help sites streamline their purchasing activities and carry out "reverse auctions," in which vendors competitively bid down their prices for goods and services to win contracts at NNSA sites.

- Individual sites from the 8 we reviewed have undertaken efforts on their own to streamline procurement and reduce costs. In addition to agreements they negotiated on their own, M&O contractor officials at 7 of the sites said they purchased goods or services through agreements negotiated by their M&Os' parent organizations to obtain better pricing. For instance, contractor officials from Lawrence Berkeley National Laboratory told us the site expanded its employees' access to scientific journals significantly at no additional cost by using the University of California's subscriptions to those journals. Officials at Pacific Northwest National Laboratory said their site leverages its M&O's parent organization's agreements with airlines, rental car companies, and banks to obtain better pricing on airfares, rental cars, and purchase cards. A senior official at that site also said, however,

---

[19]For additional information on strategic sourcing, see GAO, *Best Practices: Using Spend Analysis to Help Agencies Take a More Strategic Approach to Procurement*, GAO-04-870 (Washington, D.C.: Sept. 16, 2004).

GAO-12-255 Support Functions at DOE Sites

the site uses agreements negotiated by DOE's Integrated Contractor Purchasing Team for some purchases. In addition, contractor officials at 6 of the sites told us they made their procurement process more efficient by switching from paper-based procurement transactions to electronic ones, including electronic catalogs, which allow their sites' employees to make frequent small-value purchases without involving the site's procurement organization.

To streamline human resources, sites from the 8 we reviewed merged certain human resource functions to reduce administrative costs, according to M&O contractor officials, as well as automated some human resource services and reduced contractor employee benefits. For example:

- M&O contractor officials at Los Alamos and Lawrence Livermore national laboratories said they took steps to merge certain services, mainly in human resources. For example, Los Alamos National Laboratory provides employee benefits accounting and other services for both sites, according to the officials. Similarly, contractor officials at Y-12 National Security Complex—an NNSA site that manufactures uranium and other components for nuclear weapons—told us they took steps to merge their site's hiring, compensation, benefits administration, and other services with NNSA's Pantex Plant. According to the officials and documents describing the effort, the sites now use a common approach to planning and providing the services.

- Contractor officials from the 8 NNSA and Science sites told us they automated various human resource activities, such as hiring and training of employees or managing employee records. Officials told us that automation has helped their sites reduce the time needed to carry out the activities and, in some cases, directly lowered their costs. Contractor officials at the Y-12 National Security Complex told us their site's automation of its employee records eliminated the time-consuming management of paper records. Contractor officials at Lawrence Livermore and Los Alamos national laboratories said they moved more of their employee training online, reducing the travel or other costs of providing in-person training. Contractor officials at some sites told us that automation of such functions became necessary after, or in anticipation of, significant cuts in the numbers of support staff, including at one site where officials said over one-third of staff for some support functions were cut to address a budget shortfall. In other cases, contractor officials said that their sites automated human resource activities after evaluating the efficiency and effectiveness of existing policies and processes. Contractor officials at Lawrence Berkeley National Laboratory reported they took steps to standardize and automate the site's hiring, after an evaluation revealed that the laboratory's 14 scientific and 6 support divisions had widely varying policies and processes, requiring hiring decision approval from one official or as many as seven different officials. According to these officials, the laboratory's automated tool helped standardize hiring processes and reduce the amount of time for bringing on new staff.

- To reduce sites' costs for providing health and pension benefits to contractor employees and retirees, all NNSA and Science sites we reviewed reported making reductions in their benefit programs. In the area of health benefits, contractor officials at the 8 sites told us they increased the share of health benefit costs that contractor employees or retirees pay, for instance, by narrowing the number of health plan options and moving to plans with higher out-of-pocket expenses. Contractor officials at 3 of the sites said they eliminated postretirement medical benefits for newly hired contractor employees, and at two sites began requiring such employees to obtain their prescription medications from lower-cost mail-order pharmacies. In the area of pensions, contractor officials at 6 of the 8 sites said they made changes to their plans, reducing the overall amount of future pension benefits current contractor employees may receive in retirement. For instance, contractor officials from 3 of the sites told us they closed their traditional defined-benefit pension plans to newly hired contractor employees, and some sites made changes affecting future benefits to current employees' pensions. According to documents, Sandia National Laboratories changed its pension formula to reduce the amount of the annual retirement benefit that current contractor employees will be eligible to receive on future earnings. Another site, Brookhaven National Laboratory, reduced, from 10 percent to 9 percent of employee salaries, the amount it contributes to its newly hired contractor employees' defined-contribution pensions, according to documents.

To streamline facilities and infrastructure and reduce costs, DOE and contractors have undertaken various efforts to reduce the number of facilities and the amount of space they must maintain, lower the maintenance and operating costs of existing buildings and space, and improve how sites modernize and upgrade their facilities and infrastructure. For example:

- In 2007, Science adopted a more centralized approach to upgrading facilities and infrastructure at its laboratories to reduce costs and improve the quality of the research facilities. According to documents outlining this approach and Science officials, many facilities at Science's sites are aging, have grown more costly to maintain, and cannot easily support the equipment needed for modern scientific research, which may require environments free from vibrations or other special conditions. Until recently, Science's M&O contractors have largely been responsible for prioritizing and funding major improvements to sites' facilities and infrastructure, according to Science officials. To better leverage these funds, however, Science implemented a new approach, in which a centrally managed process is used to prioritize funding for modernizing facilities and infrastructure at all 10 Science sites, in many cases replacing the fragmented approach used previously. According to Science officials, the new approach has helped Science tie modernization efforts more closely to mission needs while bringing down the costs and lead times of these efforts. In one such effort, Oak Ridge National Laboratory was able to replace two outdated research facilities with a modern laboratory building at a lower cost and more quickly than would have been feasible under the former approach. Construction of the new building took about 2 years and was completed in 2011 for approximately $96 million. This building was completed more quickly and at less than half the cost of renovating the outdated facilities, which, according to Science officials, was the most likely modernization option under the previous funding approach. Also, according to project documentation, estimated energy costs for the new building are 40 percent less than for the renovated facilities.

- Individual sites also took various steps to streamline facilities and infrastructure. Officials at the 8 NNSA and Science sites we reviewed told us that they had consolidated staff and equipment into less space, reducing the costs of maintaining and operating space at their sites. In some cases, sites were able to repurpose the vacated space or demolish buildings to reduce the sites' overall footprint. In other cases, sites vacated inactive nuclear facilities—which can be costly to maintain, even if not inhabited—to prepare for their eventual cleanup and removal. Contractor officials at 3 of the 8 sites also told us they made better use of existing work space by changing how that space is allocated to staff and improved the quality of facility and infrastructure services through improved strategic planning for facilities and by changing how these services are organized and carried out. For example, contractor officials from Pacific Northwest National Laboratory reported that in 2011 they began relocating infrequently used laboratory equipment to a central storage facility, freeing up an additional 7,000 square feet of work space. In addition, some sites took steps to lower their energy costs. Officials at 3 of the sites we reviewed said they did so by negotiating with their utilities for better rates. Through contract negotiations between DOE and New York state, for instance, Brookhaven National Laboratory significantly lowered its utility costs by purchasing excess hydropower from New York state through a local utility, according to documents describing the effort and laboratory officials. Similarly, contractor officials at Oak Ridge National Laboratory said DOE negotiated with the utility for a less expensive rate that applies to manufacturing facilities, lowering electricity costs for the lab and adjacent DOE facilities.

Streamlining efforts described by DOE and officials at the NNSA and Science sites we reviewed appeared to incorporate many of the key practices for streamlining and improving efficiency in federal programs and functions identified in our September 2011 report.[20] These key practices include examining the efficiency and effectiveness of organizational structures and processes, targeting both short-term and long-term efficiency gains, and building capacity for further streamlining. Some of the efforts implemented by DOE and M&O contractors at the 8 sites involved Lean Six Sigma,[21] or targeted short-term and long-term efficiency gains. To build capacity for further streamlining, NNSA created a Business Management Advisory Council in 2009 to improve collaboration among its sites and encourage continuous streamlining of sites' support functions. Likewise, contractor officials at individual sites said processes are in place to promote continuous improvement at their sites. For example, a contractor official at the Y-12 National Security Complex said the site's productivity improvement initiative encourages staff to identify streamlining and cost-savings opportunities for mission activities and support functions. In fiscal year 2010, the site identified 244 streamlining initiatives—ranging from improvements to how the site's vehicle fleet is managed to the creation of an apprenticeship program for

---

[20]GAO, *Streamlining Government: Key Practices from Select Efficiency Initiatives Should Be Shared Governmentwide*, GAO-11-908 (Washington, D.C.: Sept. 30, 2011).

[21]Lean Six Sigma is a data-driven approach used in the private sector and government for analyzing work processes based on the idea of eliminating defects and errors that contribute to losses of time, money, opportunities, or business. See GAO-11-908.

training new facility maintenance workers—which were tracked internally and shared with NNSA.

Additional Streamlining Opportunities

DOE and contractor officials identified opportunities to expand existing streamlining efforts to additional sites and undertake new efforts in procurement, human resources, facilities and infrastructure, and other support functions, such as information technology. The DOE and contractor officials also noted opportunities in other support functions, including DOE-led opportunities involving multiple sites and opportunities considered by individual sites from among the 8 NNSA and Science sites we reviewed. For example:

- To further streamline procurement, the Deputy Secretary of Energy called on DOE organizations and M&O contractors to expand strategic sourcing to leverage DOE's buying power more effectively and achieve significant cost savings. In an August 2010 memorandum, the Deputy Secretary noted that successful expansion would require close collaboration among DOE and its M&O contractors and cited NNSA's central Supply Chain Management Center as a possible model for other organizations in the department. Since this memorandum was issued, other organizations in DOE, including Science, have been evaluating options for expanding strategic sourcing at the sites they oversee. In addition, individual sites among the 8 we reviewed identified further opportunities to streamline procurement. For instance, according to contractor officials, Oak Ridge National Laboratory plans to increase its use of strategic sourcing agreements and, by 2014, reduce by half the time needed to conduct its site's procurements.

- To further streamline human resources, NNSA's Business Management Advisory Council is considering whether to consolidate human resources and other services, such as payroll and finance, at all NNSA sites. In a March 2011 white paper, NNSA concluded that a centralized approach to providing these services is technically feasible and could lead to cost savings, but it would require further study of potential barriers, such as the need to standardize information systems and work processes across NNSA's M&O contractors. Individual sites also identified further streamlining opportunities. A contractor official at 1 site said the site was considering outsourcing certain human resource functions, such as employee counseling services. In the area of retirement benefits, contractor officials at 2 of the 8 sites said their sites were considering closing postretirement medical benefit programs to new contractor employees, while officials at another site said they were considering changing their defined-benefit pension plan to reduce the amount of pension benefits new employees may be eligible to receive.

- To further streamline facilities and infrastructure, both NNSA and Science officials identified opportunities to improve facilities and infrastructure planning and reduce infrastructure, energy, and other costs at multiple sites. In December 2008, NNSA selected its preferred approach for transforming its nuclear weapons sites—including sites' facilities and infrastructures—which it outlined in an October 2008 Environmental Impact Statement.[22] In particular, special nuclear material, including plutonium and highly enriched uranium, would be consolidated into fewer locations among NNSA sites, enabling sites to reduce security, infrastructure, or other costs associated with storing and safeguarding the nuclear material. At least one site, Lawrence Livermore National Laboratory, plans to remove significant stockpiles of special nuclear material from the site by the end of fiscal year 2012 and, according to M&O contractor officials there, reduce the size and cost of its security forces. Under a separate effort, NNSA is considering adopting a more centralized and strategic approach to modernizing its sites' facilities and infrastructure, as well as removing unneeded buildings and reducing sites' footprints. According to NNSA officials who oversee facilities and infrastructure, a more centralized and strategic approach is needed to ensure that improvement efforts are more closely tied to long-term mission needs. To further streamline at Science's sites, Science and its contractors determined that additional collaboration among sites could help DOE achieve governmentwide energy efficiency requirements, set by the President in Executive Order 13514.[23] In a January 2011 proposal,

---

[22]See *Record of Decision for the Complex Transformation Supplemental Programmatic Environmental Impact Statement—Operations Involving Plutonium, Uranium, and the Assembly and Disassembly of Nuclear Weapons*, Dec. 19, 2008 (73 FR 77644); *Record of Decision for the Complex Transformation Supplemental Programmatic Environmental Impact Statement—Tritium Research and Development, Flight Test Operations, and Major Environmental Test Facilities*, Dec. 19, 2008 (73 FR 77656); and *Complex Transformation Supplemental Programmatic Environmental Impact Statement*, Oct. 24, 2008 (DOE/EIS-0236-S4).

[23]Exec. Order No. 13514 (Oct. 5, 2009), 74 Fed. Reg. 52117, *Federal Leadership in Environmental, Energy, and Economic Performance*, directs federal agencies to set energy efficiency goals for their facilities and operations, including specific targets for reducing their production of greenhouse gases, such as carbon dioxide.

Science and contractor officials identified various approaches that Science's 10 laboratories could take on their own or collaboratively to improve DOE's overall energy efficiency.

In addition, officials at NNSA and Science sites reported that there were opportunities to streamline and reduce fragmentation in other support functions, such as information technology, safety, and security. Proposals for additional streamlining in these support functions involved DOE-led efforts at multiple sites, as well as opportunities identified by contractor officials at individual NNSA and Science sites from the 8 we reviewed. For example, NNSA is in the early planning stages of an organizationwide effort to upgrade its sites' information technologies and improve the security of the sites' networks, according to NNSA officials. Under a key component of this effort, wireless technologies would be used to remotely monitor potentially dangerous conditions in nuclear environments, helping contractors improve safety and reduce costs by shortening the amount of time workers need to spend in these environments, according to DOE documents. NNSA is currently testing the feasibility of using wireless technologies at Livermore and Los Alamos national laboratories; another site, Sandia National Laboratories, has already been using these technologies to improve the safety and efficiency of operations. Under another key component, NNSA sites' current patchwork of network technologies and architectures would be replaced with standardized technologies and architectures, reducing the need for site support personnel and improving NNSA's ability to secure its sites' networks, according to NNSA officials.[24] A related component would reduce the need for technology and support personnel at sites by relocating a number of site-supported computer applications to a "cloud" computer—an emerging model in which computer services are provided centrally through the Internet.[25] The officials also reported that providing wireless technologies and upgrading sites' networks through a coordinated effort would allow NNSA to leverage its purchasing power more effectively than through fragmented efforts at sites. In addition, contractor officials at 3 of the Science sites we reviewed also told us they were considering cloud

---

[24]For more information on network architecture, see GAO, *Organizational Transformation: A Framework for Assessing and Improving Enterprise Architecture Management (Version 2.0)*, GAO-10-846G (Washington, D.C.: Aug. 5, 2010).

[25]For more information, see GAO, *Information Security: Additional Guidance Needed to Address Cloud Computing Concerns*, GAO-12-130T (Washington, D.C.: Oct. 5, 2011).

GAO-12-255 Support Functions at DOE Sites

computing for their sites. Officials at the 4th Science site, Lawrence Berkeley National Laboratory, said their site has been using cloud computing since 2010 for certain tasks, such as employees' e-mail and calendar, which an offsite vendor remotely manages for the laboratory.

Challenges Could Hinder Additional Streamlining Efforts

DOE and contractor officials also cited a variety of challenges that could hinder implementation of additional streamlining efforts. One such challenge is the potential difficulty of getting M&O contractors to coordinate and adopt a more centralized, and less fragmented, approach. DOE and contractor officials told us that in some cases, the sites were reluctant to adopt a centralized approach to providing support functions because such an approach may not always be more effective. For example, in response to the Deputy Secretary's August 2010 memo calling for an expansion of strategic sourcing, Science expressed reluctance to implement a more centralized approach to procurement, citing efficiencies of its sites' current procurement approach. In July 2011, Science reported that after assessing options for centralizing sites' procurement—including options for joining NNSA's central Supply Chain Management Center or establishing a center of its own—it concluded that the benefits of these options would not exceed the costs. Despite its less centralized approach, Science concluded that its sites already benefit from key aspects of strategic sourcing—including analysis of sites' spending and leveraged buying through contractors' parent organizations and DOE's Integrated Contractor Purchasing Team—and have achieved significant and potentially greater savings through their own procurement activities. In a September 2011 study, an outside consultant hired by Science also concluded that Science's laboratories would not achieve further cost savings by joining NNSA's central Supply Chain Management Center but could benefit from methods used there.[26]

[26]The study—"Use of Supply Chain Management Center by Department of Energy Office of Science National Laboratories," McCallum-Turner, Sept. 16, 2011—compared procurement cost savings in fiscal year 2010 between Science laboratories and NNSA's Supply Chain Management Center. The report concluded that joining with NNSA's center would not benefit Science, because "major" Science laboratories are achieving greater savings on their own. Other Science laboratories, however, which are achieving comparable or less savings, could benefit from some of the tools used by NNSA's center. The report also, however, identified potential limitations to its analysis, including difficulty with comparing data from multiple sites that may not be comparable.

We did not evaluate Science sites' savings potential under a more centralized approach to procurement, but procurement officials in DOE told us that Science likely has additional opportunities to leverage its laboratories' buying power, because Science has not comprehensively analyzed its sites' spending. Moreover, DOE and contractor officials said that barriers to adopting a more centralized approach can potentially be overcome, despite DOE and sites' long-standing fragmentation. For example, according to NNSA's procurement director and a senior contractor official at NNSA's central Supply Chain Management Center, NNSA's M&O contractors were initially reluctant to participate in the supply chain center because they were concerned about losing their autonomy over procurement activities. In addition, because DOE competes its M&O contracts, they were reluctant to release potentially sensitive information about those activities. Officials said they were able to address these issues, in part by securing the NNSA Administrator's support for the new center and by requiring contractors to participate in the effort. Similarly, contractor officials at Los Alamos National Laboratory and Y-12 National Security Complex told us that their respective efforts to merge certain human resource activities with those of other NNSA sites were helped by the fact that the contractors involved in the efforts have common parent organizations.

In addition to the potential difficulty of getting contractors to coordinate and adopt a more unified approach, DOE and contractor officials cited other challenges to further streamlining, including the long lead times or high up-front costs that can sometimes precede cost savings or other streamlining benefits. Officials from Sandia National Laboratories, for example, told us the anticipated cost savings from eliminating new contractor employees' postretirement health benefits in 2009 would not materialize for many years because the affected employees are years from retirement. Similarly, in the March 2011 white paper evaluating options to merge sites' human resource support services, NNSA cited potentially long lead times and high expected up-front costs, which it estimated could range from $500 million to $1 billion, before the anticipated cost savings and other benefits would result. Also, according to contractor officials or documentation from several sites, efforts to automate support functions and improve facilities and infrastructure can require long lead times or costly investments up front but can also save on costs down the road, because they improve efficiency or require fewer staff.

In addition, according to DOE and contractor officials, while reducing employee benefits or laying off staff can reduce costs, these steps could also hinder contractors' ability to recruit and retain a high-quality workforce. Officials said that such steps can affect morale at the site and make it more difficult to recruit scientists, engineers, and other highly trained staff. They can also exacerbate existing challenges related to the geographic isolation of some NNSA and Science sites and competition from other industries. Officials at Sandia National Laboratories, for example, said that they had to achieve dramatic reductions in the site's pension liability while ensuring the site retained adequate numbers of experienced staff for its highly technical and specialized nuclear weapons-related workload, which they said would likely increase in the coming years. Other officials told us, however, that reductions in employee benefits may not always have adverse effects on sites' ability to recruit and retain a high-quality workforce because some benefits offered at NNSA and Science sites, such as defined-benefit pension plans, may be more generous than those offered by competing industries.

Difficulties Exist in Quantifying Cost Savings because Guidance for Estimating Savings Is Unclear and Methods Used Vary

DOE and its M&O contractors at NNSA and Science sites have estimated savings for some of the streamlining efforts undertaken since fiscal year 2007. It is difficult, however, to compare savings or quantify total savings across sites because DOE's guidance for estimating savings is unclear and the methods used to estimate savings vary. Several streamlining efforts for which DOE and M&O contractors estimated cost savings were in the area of procurement. Examples of estimated savings reported by DOE or its contractors included the following:

- According to NNSA documentation, NNSA estimated that in fiscal year 2011 its seven sites saved $106.6 million on purchases using its central Supply Chain Management Center.

- Individual sites also reported savings. For example, contractor officials at Pacific Northwest National Laboratory estimated that in fiscal year 2010 the lab saved about $35 million in procurement costs through its direct negotiations with suppliers, purchases through agreements negotiated by its parent organization, and other means.

DOE's guidance for how to estimate procurement savings, however, is not consistent and clear in all cases. As a result, the methods used to estimate these cost savings can vary widely across sites. For some procurements, standardized methods for estimating cost savings are used. For example, guidance by NNSA's central Supply Chain Management Center specifies that NNSA's seven sites use the center's preferred method for estimating cost savings from purchases made through the center, in which the price paid for goods or services is subtracted from the previous price paid. If no previous price is available or would not provide a useful comparison, the guidance identifies two other approved methods. Similarly, for purchases made through DOE's Integrated Contractor Purchasing Team agreements, the team estimates cost savings by subtracting the price paid through its agreements from the General Services Administration (GSA) price,[27] if available, or the vendor's list price, according to the contractor official who coordinates the team's activities.

For other procurements, DOE has not historically had guidance for estimating cost savings. In September 2011, to help improve consistency across the department, and comply with a governmentwide effort to measure procurement cost savings, DOE issued guidance outlining approved methods for calculating these savings. The guidance was one part of DOE's efforts to revise the department's metrics for evaluating the performance of its M&O contractors' procurement functions, in response to recommendations from M&O contractors.[28] M&O contractor representatives from DOE's newly formed Contractor Integrated Supply Chain Management Council identified the need for greater consistency in reporting of sites' procurement cost savings. Also, according to an official from DOE's Office of Procurement and Assistance Management, DOE's effort to develop guidance was responding to Office of Management and Budget (OMB) direction that agencies governmentwide, including DOE, report on their procurement cost savings. The council suggested various methods for estimating procurement cost savings, which it recommended DOE adopt as acceptable methods. These methods include determining procurement cost savings by subtracting the price that a site paid for goods and services from a comparison price, such as the previous price paid at the site or an independent estimate. DOE incorporated these methods into a new performance metric for evaluating M&O contractors'

---

[27]This price is from GSA's Multiple Award Schedule program, in which GSA negotiates contracts with a variety of vendors and allows federal agencies and contractors, including DOE's M&O contractors, to make purchases at the negotiated prices.

[28]Contractors report procurement performance to DOE annually using the Contractor Purchasing System Balanced Scorecard. The balanced scorecard requires contractors to report on a variety of procurement metrics, which DOE then uses to oversee the contractors' procurement functions.

procurement functions, which will require M&O contractors to report their annual procurement cost savings to DOE starting in fiscal year 2012.

The guidance for the new metric, however, does not specify when it is appropriate to use each method and consequently could lead to wide variations in the cost savings reported by contractors. The guidance also does not identify a preferred method for estimating cost savings; rather, it allows contractors to use any one of the methods recommended by the Contractor Integrated Supply Chain Management Council as well as any other method approved by local DOE site offices. A variety of methods may be appropriate for calculating procurement cost savings, but we found that different methods could lead to wide variations in estimated cost savings, making it difficult for DOE to oversee contractors' performance in streamlining and reducing the costs of site procurement functions. For example, the official who oversees procurement at Pacific Northwest National Laboratory said his division used the site's preferred method to estimate the cost savings resulting from negotiations with a vendor over the price of a new software package. Using this method—in which the price paid was subtracted from the vendor's opening offer—the official cited a savings of $9 million, which he considered to be an accurate estimate of the savings. According to the official, however, other, less conservative methods for estimating procurement cost savings—including those used elsewhere in DOE—would have led to higher savings estimates, including savings as high as $35 million (see fig. 2).

Figure 2: Example of the Effect of Using Different Cost Savings Estimation Methods for a 2010 Software Purchase at Pacific Northwest National Laboratory (PNNL)

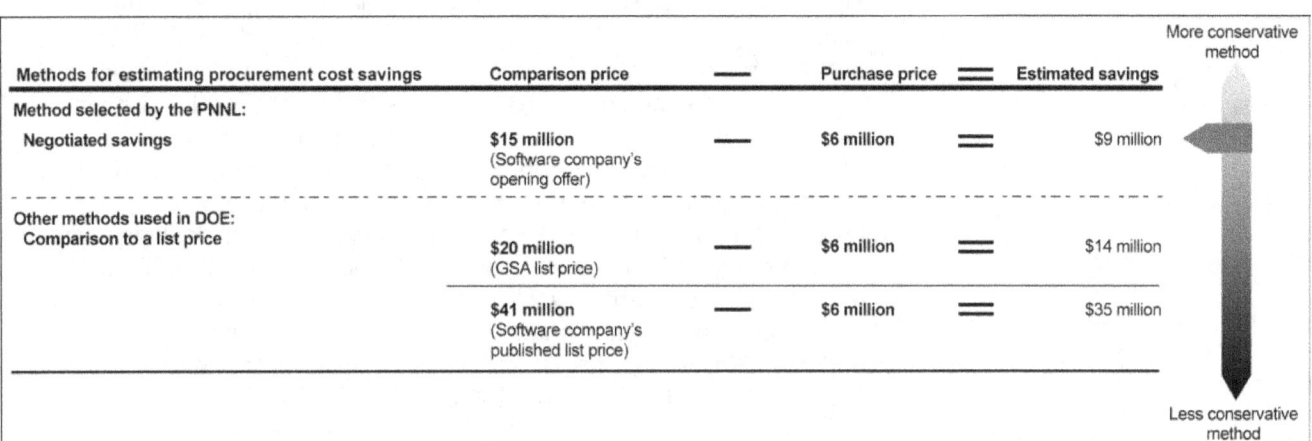

Source: GAO analysis of PNNL data.

According to an official from DOE's Office of Procurement and Assistance Management, the new guidance provides a first step to addressing these inconsistencies, but DOE expects to further clarify the guidance. He also noted that, despite the governmentwide requirement to report procurement cost savings, DOE developed the guidance with little specific instruction from OMB on the appropriate methods for measuring procurement cost savings.[29]

For cost savings estimates for support functions other than procurement, we also found that there is considerable flexibility in how savings may be calculated, which could lead to wide variations in savings estimates. For example, a contractor official at Brookhaven National Laboratory reported that the estimated savings during the first few months of the site's agreement with its utility to purchase hydropower varied, depending on the estimation method. The estimated savings totaled either $1 million or

---

[29]In November 2011, we reported that OMB guidance issued as part of its effort to reduce procurement spending and increase agencies' reporting of procurement cost savings was broad and led to inconsistent interpretations by agencies as to what constituted savings. We therefore recommended that OMB's Office of Federal Procurement Policy clarify its guidance related to measuring procurement cost savings. See GAO, *Federal Contracting: OMB's Acquisition Savings Initiative Had Results, but Improvements Needed*, GAO-12-57 (Washington, D.C.: Nov. 15, 2011).

GAO-12-255 Support Functions at DOE Sites

$7 million, depending—respectively—on whether the cost of hydropower was compared with the site's previous price for electric power or the utility's market rate. Similarly, contractor officials from Los Alamos National Laboratory reported saving $200,000 annually in travel-related costs by encouraging site employees to use video teleconferencing in lieu of attending meetings or training sessions in person. The laboratory officials told us that they could have included other cost savings in their estimate—such as costs (productivity losses) avoided when staff spend more time on their regular duties rather than travel offsite—but wanted a more conservative savings estimate. In contrast, in their savings estimate for another type of process improvement, contractor officials at the Y-12 National Security Complex included savings from needing fewer staff-hours to complete tasks. In this case, the site automated its process for counting its inventory of personal property, such as computers, which officials estimated saved $1 million in fiscal year 2011.

DOE guidance relevant for cost savings in other support functions does not clearly define appropriate methods for estimating cost savings. DOE's acquisition regulations include provisions that give M&O contractors an opportunity to earn a share of the cost savings, paid out as fee, resulting from streamlining efforts.[30] Guidance in the regulations outlines the process for identifying and verifying savings but does not clearly define appropriate methods for estimating the savings in order to award additional fee. In May 2011, NNSA revised these provisions and incorporated them into its sites' M&O contracts. The revised version defines more clearly what types of streamlining efforts may result in additional fee for contractors—specifically, efforts that reduce sites' bottom-line operating costs—but it does not define estimation methods for cost savings.[31] Instead, both the original and revised versions of the guidance allow the contractor to use any estimation methods approved by the local site office, which could lead to wide variations in savings estimates for similar efforts and make it more difficult to oversee contractors' performance in streamlining and reducing support function costs.

---

[30]Department of Energy Acquisition Regulation (DEAR) 48 C.F.R. § 970.5215-4 Cost Reduction.

[31]NNSA Deviation to DEAR 48 C.F.R. § 970.5215-4 Cost Reduction, authorized pursuant to the authority of FAR 48 C.F.R. § 1.404.

GAO-12-255 Support Functions at DOE Sites

Absent clear DOE guidance, at least one site has developed its own instructions to standardize savings estimates from its streamlining efforts. The Y-12 National Security Complex identified specific methods that site employees should use to estimate cost savings under its sitewide productivity improvement initiative. The instructions outline which elements should be used to estimate the cost savings and, because some savings span multiple years, the time period for recognizing efforts' cost savings. Furthermore, an online system is used at the site to track streamlining efforts and includes a template for estimating cost savings, used to help ensure that savings are being calculated consistently. Savings estimates are reviewed at the site and, for efforts resulting in significant savings, by local NNSA officials. If applied consistently, this guidance would standardize estimates within the Y-12 site but not other sites.

Given the uncertainty of measuring cost savings from streamlining, DOE and site officials told us they do not always measure cost savings. For example, contractor officials at Brookhaven National Laboratory told us they took a number of steps to improve the efficiency of processes at the site—including streamlining the process for admitting visitors and guests to the laboratory—but did not attempt to measure the cost savings because the process improvements did not reduce the laboratory's bottom-line operating costs. Even when streamlining efforts may result in reductions to bottom-line costs, however, anticipated savings can be difficult to estimate accurately. For example, some contractor officials told us that, in order to estimate savings from changes to contractor employees' pension or health care benefits, they often had to predict future events—such as the long-term performance of financial markets or employees' expected use of health care services. Sandia National Laboratories expects recent changes to the site's pension plan will save approximately $380 million from 2011 to 2020, but the actual amount saved will depend, in part, on the future performance of the plan's underlying financial assets, which can be difficult to predict.

Conclusions

DOE spends over $5 billion dollars each year on support functions provided by M&O contractors at NNSA and Science sites. Growing federal deficits and increasingly uncertain future federal budgets have necessitated that M&O contractors evaluate areas that could be streamlined or provide cost savings in support costs at these sites, thereby maximizing funds available for the sites' national security, research, and energy development missions. DOE also has actively sought opportunities and implemented measures to more effectively use federal funds.

Moving forward, DOE's ability to oversee and facilitate additional streamlining of support functions provided by M&O contractors will require complete and comparable data on support function costs. DOE has implemented an improved reporting system for support costs. Moreover, DOE has partially implemented a quality control process to begin to ensure that data in the new cost reporting system are complete and comparable. DOE has not yet completed implementing its quality control process, however, and the completion date has now slipped from the end of September 2011 to sometime in 2012. Key steps, such as peer reviews, are not yet used. Until a quality control system is completely implemented, DOE cannot have full confidence that the support cost data it collects and uses to oversee contractor performance are complete, accurate, and reliable.

In addition to improving support cost data, DOE and its contractors have made substantial efforts to streamline support functions and reduce costs. Also, they have identified additional opportunities to expand these efforts and implement new approaches. Streamlining efforts undertaken at some sites may be appropriate at other NNSA or Science sites, and there may be other opportunities to streamline or reduce fragmentation that could be pilot-tested and implemented at sites. Some of these opportunities will require close collaboration between DOE and its M&O contractors to reduce the effects of long-standing fragmentation of site management. DOE has begun taking active steps to reduce the effects of fragmentation, such as NNSA's creation of its Business Management Advisory Council to improve collaboration among its sites and encourage continuous streamlining of sites' support functions. However, as DOE and contractor officials have pointed out, barriers to collaboration and other challenges could hinder further streamlining.

DOE has also undertaken efforts to better standardize its guidance on the appropriate methods to use when estimating cost savings. Because its guidance gives contractors considerable flexibility in choosing the appropriate methods for estimating cost savings, its estimates could still vary widely. Consequently, DOE's ability to determine whether contractors' efforts to further streamline costs are effective is limited.

Recommendations for Executive Action

To help reduce support costs or make more effective use of DOE and contractor resources, as well as to improve oversight of M&O contractors' support functions at NNSA and Science sites, we recommend the Secretary of Energy take—or, as appropriate, direct the Administrator of NNSA and the Director of the Office of Science to take—the following three actions:

- fully implement a quality control system for DOE's institutional cost system, including steps such as peer reviews, to ensure that data collected and used by DOE on support function costs are complete and comparable for monitoring sites' support functions;

- assess whether all appropriate efforts to streamline DOE support functions or reduce support function costs are being taken at NNSA and the Science sites and ensure that necessary steps are taken to address challenges limiting implementation of cost savings efforts; and

- clarify DOE's guidance on the preferred methods to use for estimating cost savings, including under what circumstances each method should be used, to ensure more consistency in how cost savings are estimated for various streamlining efforts and a more comparable assessment of results.

Agency Comments and Our Evaluation

We provided a draft of this report to DOE for its review and comment. In written comments, NNSA's Associate Administrator for Management and Budget, responding on behalf of DOE and NNSA, wrote that DOE agreed with the report's findings and three recommendations. DOE provided additional information about its planned actions for addressing the recommendations that included implementing a peer review process to ensure the quality of its support cost data, establishing an organization in Science comparable to NNSA's Business Management Advisory Council to verify that all appropriate streamlining opportunities are taken and challenges addressed, and clarifying the department's recent guidance on the methods for estimating cost savings. DOE's written comments on our draft report are included in appendix I. DOE also provided technical comments, which we incorporated into the report as appropriate.

We are sending copies of this report to the appropriate congressional committees, the Secretary of Energy, the Administrator of NNSA, and other interested parties. The report also is available at no charge on the GAO website at http://www.gao.gov.

If you or your staff members have any questions about this report, please contact me at (202) 512-3841 or aloisee@gao.gov. Contact points for our Offices of Congressional Relations and Public Affairs may be found on the last page of this report. GAO staff who made major contributions to this report are listed in appendix II.

Gene Aloise
Director, Natural Resources
    and Environment

Appendix I: Comments from the Department of Energy

Department of Energy
National Nuclear Security Administration
Washington, DC 20585

January 13, 2012

Mr. Gene Aloise
Director
Natural Resources and Environment
Government Accountability Office
Washington, DC 20548

Dear Mr. Aloise:

The National Nuclear Security Administration (NNSA) and the Department of Energy (DOE) appreciate the opportunity to review the Government Accountability Office's (GAO) draft report, GAO-12-255, *DEPARTMENT OF ENERGY: Additional Opportunities Exist to Streamline Support Functions at NNSA and Office of Science Sites.* I am pleased to provide a consolidated response to that report on behalf on both NNSA and DOE.

In general, we believe the GAO did a commendable job in tackling a very broad and challenging area and appreciate the balanced presentation of the many initiatives and accomplishments we have achieved in pursuing cost savings and leveraging those savings where appropriate across DOE and NNSA laboratories. To that end, we generally agree with the findings and recommendations contained in the report and will continue to develop and implement corrective actions to further improve our ability to accurately estimate cost savings from our initiatives and demonstrate our commitment to effective stewardship of taxpayer dollars.

I have enclosed a summary of our planned actions to address the three recommendations noted in the report. In addition, while the auditors have done a thorough job, I have also included general comments to further clarify certain salient points in the report and/or improve the factual accuracy in selected areas.

If you have any questions related to this response, please contact Dean Childs, Director, Office of Management Controls and Assurance, at 301-903-1341.

Sincerely,

Cynthia Borstar-ffa

Kenneth W. Powers
Associate Administrator
for Management and Budget

Enclosure

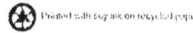

Printed with soy ink on recycled paper

2

National Nuclear Security Administration and Department of Energy
Consolidated Comments on the GAO Draft Report
"Department of Energy: Additional Opportunities Exist to Streamline
Support Functions at NNSA and Office of Science Facilities," GAO-12-255

Initial Response to Report Recommendations

Recommendation 1
"Fully implement a quality control system for DOE's institutional cost system, including steps such as peer reviews, to ensure that data collected and used by DOE on support function costs are complete and comparable for monitoring sites' support functions."

Management Response: Agree. As noted in the report, the Department and NNSA (lead by the DOE Office of Chief Financial Officer) plan to implement a "Peer Review" process to ensure quality and consistency in the support costs reported across all sites. This process is being refined and is planned to be implemented in fiscal year 2012.

Recommendation 2
"Assess whether all appropriate efforts to streamline DOE support functions or reduce support function costs are being taken at NNSA and the Office of Science sites and ensure that necessary steps are taken to address challenges limiting implementation of costs-savings efforts."

Management Response: Agree. NNSA and the DOE Office of Science (SC) have and will continue to work with our laboratories to find practical and cost-effective ways to streamline support functions, reduce costs, and ensure that we're addressing challenges. In NNSA, the NNSA Chief Financial Officer and the Office of Acquisition and Project Management will be working with the Business Management Advisory Council (BMAC) to help define corporate strategies to help ensure cost savings are promoted and leveraged across sites where feasible. In addition, SC recently established the Science Laboratory Operations Improvement Committee (SLOIC), an organization comparable to NNSA's BMAC, that brings together SC laboratory, site office, and headquarters personnel to track and communicate cost avoidance and cost savings information among the SC organization and for the DOE leadership, as well as to organize SC and assure in an ongoing fashion that all the SC contractors are continuing to take advantage of all appropriate mechanisms for cost avoidance and savings.

3

Recommendation 3

"Clarify DOE's guidance on preferred methods to use for estimating cost savings, including under what circumstances each method should be used, to ensure more consistency in how cost savings are estimated for various streamlining efforts and a more comparable assessment of results."

Management Response: Agree. While the Department has issued corporate guidance to identify preferred and prioritized methods for calculating cost savings, we recognize that it did not clearly specify that the methods were presented in order of preference and that guidance may be further clarified regarding the utilization of those methods. Clarifying guidance will be drafted.

Appendix II: Contact and Staff Acknowledgments

GAO Contact	Gene Aloise, (202) 512-3841 or aloisee@gao.gov
Staff Acknowledgments	In addition to the individual named above, Janet Frisch, Assistant Director; Eric Bachhuber; James Espinoza; Daniel Feehan; Allyson Goldstein; Mehrzad Nadji; Alison O'Neill; Josie Ostrander; Cheryl Peterson; Jeff Rueckhaus; Michael Silver; and Vasiliki Theodoropoulos made key contributions to this report.

GAO's Mission	The Government Accountability Office, the audit, evaluation, and investigative arm of Congress, exists to support Congress in meeting its constitutional responsibilities and to help improve the performance and accountability of the federal government for the American people. GAO examines the use of public funds; evaluates federal programs and policies; and provides analyses, recommendations, and other assistance to help Congress make informed oversight, policy, and funding decisions. GAO's commitment to good government is reflected in its core values of accountability, integrity, and reliability.
Obtaining Copies of GAO Reports and Testimony	The fastest and easiest way to obtain copies of GAO documents at no cost is through GAO's website (www.gao.gov). Each weekday afternoon, GAO posts on its website newly released reports, testimony, and correspondence. To have GAO e-mail you a list of newly posted products, go to www.gao.gov and select "E-mail Updates."
Order by Phone	The price of each GAO publication reflects GAO's actual cost of production and distribution and depends on the number of pages in the publication and whether the publication is printed in color or black and white. Pricing and ordering information is posted on GAO's website, http://www.gao.gov/ordering.htm. Place orders by calling (202) 512-6000, toll free (866) 801-7077, or TDD (202) 512-2537. Orders may be paid for using American Express, Discover Card, MasterCard, Visa, check, or money order. Call for additional information.
Connect with GAO	Connect with GAO on Facebook, Flickr, Twitter, and YouTube. Subscribe to our RSS Feeds or E-mail Updates. Listen to our Podcasts. Visit GAO on the web at www.gao.gov.
To Report Fraud, Waste, and Abuse in Federal Programs	Contact: Website: www.gao.gov/fraudnet/fraudnet.htm E-mail: fraudnet@gao.gov Automated answering system: (800) 424-5454 or (202) 512-7470
Congressional Relations	Katherine Siggerud, Managing Director, siggerudk@gao.gov, (202) 512-4400, U.S. Government Accountability Office, 441 G Street NW, Room 7125, Washington, DC 20548
Public Affairs	Chuck Young, Managing Director, youngc1@gao.gov, (202) 512-4800 U.S. Government Accountability Office, 441 G Street NW, Room 7149 Washington, DC 20548